Bibliographic information published by the German National Library:

The German National Library lists this publication in the National Bibliography; detailed bibliographic data are available on the Internet at http://dnb.dnb.de .

Imprint:

Print and binding: Books on Demand GmbH, Norderstedt Germany
ISBN: 9783668334984

This book at GRIN:

http://www.grin.com/en/e-book/342136/haphazard-property-development-and-its-consequences-a-study-in-the-bawku

Wuni Ibrahim Yahaya

Haphazard Property Development and its Consequences. A Study in the Bawku Municipality, Ghana

GRIN Publishing

Haphazard property development in the Bawku municipality: a case study of Bawku highways.

WUNI IBRAHIM YAHAYA

Department of Land Economy, Faculty of Built Environment, KNUST, Kumasi.

Abstract: the paper sought to make a paradigm shift from conceptual frameworks to empirical studies to investigate the motivations for and effects of haphazard property development in the Bawku municipality. The study adopted a case study research design and made use of both qualitative and quantitative data from the field survey and literature review. The data primary data was obtained from 120 property owners, the assemblyman of the area and the planning authorities of the Bawku municipality. The study relied on both primary and secondary data. The primary data was obtained through questionnaires, group interviews and focus group discussion and the secondary data was obtained from the literature review. The motivation for the haphazard property development was categorized into property owners' perspective and stakeholders' perspective. From the property owners' perspective, a number of motivation were identified including ignorance on planning and building regulations 90 (75.00%), nature of the property 80 (66.67%) and corruption 68 (56.67%). From the stakeholders' perspective, a number of motivations were identified and ranked including indiscipline of the property owners (1st), population growth and unregulated urbanization (2nd) and logistic and capacity gap (3rd). The paper also identified a number of effects of the haphazard property development including threat to future development, destruction of the natural environment and biodiversity and poor planning outlook. The paper therefore recommended that the concerted effort of both the planning authorities and stakeholders is needed to create awareness on the need to obtain building or planning permits before commencing property development and punitive sanctions should be meted out to offenders identified including demolition of such development at the earliest stage of development, conviction, imprisonment or convictions with a fine.

Keywords: Haphazard, property development, Bawku Highways, Bawku Municipality

Contents

1.0 INTRODUCTION

For a couple of years now, peace is now returning to the Bawku Municipality. The result has been that property development for residential and other purposes are on the sharp increase in a manner that invariably all pieces of land are now becoming a part of the built environment. This developmental motives of the folks is however unregulated. Houses, shops and other buildings are erected everywhere in the Municipal and the worst been at Bawku Highways. The seemingly unregulated nature of property development is threatening the true developmental layout and planning of the area. One will find a lot of notices nearly on all Buildings by an authoritative notice "*stop work, produce permit by the Bawku Municipal Planning Authority (BMPA*)" all over the area.

This is either a manifestation of the inability of the planning authorities to regulate property development or a shear level of indiscipline among property developers stemmed from ignorance or the refusal to obey rules and regulations. These forms of unauthorized building constructions has been known to be on the upsurge in most Towns, Cities and municipals in third world countries around the globe and Ghana is among the violent practitioners of this unhealthy planning indiscipline.

It is baffling to realize that these haphazard buildings are sited in viable lands meant for other economic uses and at worst some of the buildings are situated in areas liable to flood, buffer zones for road constructions, water courses and other disaster prone areas. Adjei (2010) rather narrowed it that these buildings were located on surplus virgin lands which were neglected by planning authorities. This is far from been right as in the case of the Bawku Highways.

In a broader view, Acquah and Harrison (2004) opined that this planning disorder is eating deep into planning status of major cities in Ghana such as Tamale, Kumasi, Accra among other emerging cities where haphazard building construction are conspicuous everywhere. Kumar (2012) rather identified a very disturbing practice where some cities demarcate certain land areas for specific developmental need yet such area are still caught in the web of unauthorized development. He noted that there was haphazard siting of land uses where agricultural lands were used for housing, residential areas on industrial lands, residential buildings on reserved lands among other disturbing allocation of land uses. The result is that these areas do not get government utilities such as electricity, water supply, roads and sewage treatment mechanisms.

Aklorbotu (2013) agrees with Kumar (2012) in the haphazard allocation of land uses when his study opined that the Secondi-Takoradi Metropolis have some of its lands reserved for future development been used for the erections of buildings and that some of the lands ear-marked for road buffer zones and other community development planning were occupied with buildings from unauthorized owners.

Nearly all the haphazardly sited buildings in the Bawku Highways do not either have planning permission, development permission and or rarely environmental permits. Some of the buildings are located too close to access routes, area liable to floods and are mostly found in unauthorized locations or unapproved places.

A number of serious problems and consequences can be noted in this haphazard development which includes the waste of economic lands as most of the lands are used for compound houses instead of high rise buildings. It also creates planning and development problems as some future developments might not find suitable lands for establishment. The result has been that it gives the area poor layout and at worst threatens the area to an urban sprawl.

It is on this motivation that this research paper seeks to undertake an empirical investigation into the causes and effects of haphazard property development in Highways of the Bawku Municipality and to identify practical mechanism to stop the practice or to minimize the impact of this unfathomable planning practice.

2.0 REVIEW OF RELEVANT LITERATURE

2.1 Factors motivating the haphazard siting of properties

The United Nations (UN) (2007) conducted a study and categorized the factors motivating haphazard property development into political, physical, institutional and socio-economic actors. A similar study conducted by Ioannidis et al (2007) also categorized the factors into broad segments of administrative, legal, economic and social factors. However, in the case of this empirical investigation in the Bawku Highways, the researcher did not make such groupings and so all the factors were studied.

The United Nations (2007) rather opined that haphazard property development could be traced to unrealistic zoning regulations and unnecessary bureaucracy in issuing building permits. Owusu also identified that haphazard property development can be associated to complex legislations and Freiku (2003) asserted a similar argument that haphazard property development could be traced to complex legislations and corrupt nature of planning officials in executing their planning responsibilities. A rather different couple of factors such as lax planning policy administration, obsolete planning laws and inability of planning officials to provide realistic building layouts were advanced by Magigi and Majani (2006) to be some of the factors that motivate haphazard property development

Melesse (2006) argued that lack of awareness of planning regulations could be a great cause of haphazard property development. This argument rather reflects the level of education or literacy in the Bawku municipality as most of the property owners were largely illiterates until recently. Somiah et al (2015) disagreed with Melesse (2006) when their study is Takoradi revealed that haphazard property developers had at least basic education and are not necessarily illiterate. Kings-Amadi (2004) rather stressed that low level of awareness on planning regulations, ignorance and lack of planning knowledge on the part of the property developer, bureaucratic and expensive planning permission acquisition process and shear indiscipline of the part of the property developers are the motivations for haphazard property development.

Cheema (1993) also opined factors which are political in nature and included non-enforcement of housing policies and unsustainable political will in initiate planning policies as some of the motivations for haphazard property development. Sietchiping (2000) attributed some of the motivations to be associated with location and nature of the land and argued strongly that the evidence exist where most of the haphazard property siting are disaster prone and virgin urban lands, very steep slopes, river banks in Kenya, Yaoundé and Bombay respectively.

Smith (1980) believes that one factor motivating haphazard property development has to with lack of detailed coordination of planning operation by the appropriate authorities. This view is also supported by Ali and Sulaiman (2006) who opined that another factor is discriminating regulations and public spending. High cost of land, scarcity of land, high rental charges and chronic poverty were a group of motivations for haphazard property development opined by Adjei (2010) and Sietchiping (2000).

2.2 Effects of Haphazard property development

Haphazard property development has been known to create discomfort for the inhabitants of over-crowded areas. It has been known to be distorting existing layout or planning schemes. Freiku (2003) also agrees with this when he identified that sometimes sites for public rights and places comprising natural reserves, roads, market, schools, sanitations, open spaces among could be used for personal property development. This is indeed detrimental to the development potentials of the areas.

In haphazardly developed areas, property owners have been known to violate building regulations thereby putting up buildings at certain height and altitude that threatens the safety of the inhabitants. Somiah et al (2015) opined that some property owners have been identified to be violating the National Building Regulation of 1996 by developing their plots to nearly 100% instead of the maximum stipulated 80%.

Haphazard property development creates disconformity in land use planning. An agricultural land might be used for industrial building and the vice versa in such areas. Adams (2014) opined that only about 24% of property development in the metropolis in Ghana correlates with land use planning and accordingly connotes that about 76% of such development in the metropolis are haphazardly sited. This indeed accounts for the urban sprawl in the metropolis in Ghana.

The creation of congenial avenues for squatters has been identified as one of the consequences of haphazard property development. This has also created chances for greater impacts from natural disasters. This was also opined by Adams (2012) when he identified that during floods most inhabitants in haphazardly developed areas suffer ill health. The results in most cases had led to series negative environmental impacts in the affected areas (Ioannidis *et al*, 2007).

Tragedy of the commons involving the blockage of safe passages meant for the general public has been identified as one of the consequences of haphazard property development. Freiku (2003) argued this better by asserting that such properties could be located in public places such 'as schools, markets, sanitation sites, open spaces, nature reserves, parks and roads without restraint' and cautioned that 'this practice if not curbed has the tendency of creating both hygiene and safety related problems'.

3.0 RESEARCH METHODOLOGY AND MATERIALS

3.1 Research Design

The haphazard property development in the Bawku Highways was investigated using a case study research design. Creswell (2007) opined that a case study becomes the most powerful research design when the researcher uses several sources of evidence in the study process.

3.2 Target Population and Sample Frame

The target population in this research paper included all the owners of property that were haphazardly sited, the planning authorities and the assemblymen in the Bawku municipality who were within the Bawku Highways jurisdiction. The property-owners were selected because they were the actual players whose buildings were haphazardly located or sited at unapproved places. The planning authorities were also selected because they are responsible for the planning responsibilities of the area and they ought to have made attempts to right the wrongs that were prevailing and the assemblymen were selected because they are the link between the community and the municipal and they aught have reported the matter.
Therefore sample frame in this regard consisted of the list of all property owners whose buildings were haphazardly located or sited at unapproved places, the development planning team, the Town and Country planning team of the Bawku Municipality and the Assemblymen within the Bawku Highways. However, the sample frame was not considered in the data collection from the property owners because there were no available records containing the list of addresses of all the property owners whose buildings were located at unapproved places.

3.3 Sample size and Sampling techniques

The sample size that was considered for study was 120 property owners, the heads of development planning department and the Town and Country planning Department as well as one Assemblyman in the Bawku Highways. The sample size for the property owners was an arbitrarily figure that was considered because of the difficulty in getting the target groups to respond to the interviews and the questionnaires. Besides most of them were not educated and it was really time consuming having to repeat a question several times before answers could be provided. The sampling techniques adopted for this study were the purposive sampling and the snowball sampling techniques. These techniques were adopted because the sample frame was not readily available. Besides, the researcher also lives in the community and knew the owners of some of such buildings.

3.4 Sources of data

The sources of data that were relied upon in this research paper were both primary and secondary data. The primary data were collected from the respondents based on the research objectives and care was taken to ensure that schedule of questions were sufficient to provide the required data. The administration of the questions and the interviews facilitated the collection of primary data and the secondary data were collected from journals, the internet, research papers, articles, theses and were accordingly cited and referenced.

3.5 Data collection method

The data collection methods that were adopted included personal interviews, self-administered questionnaires and focus group discussion. In both methods a prospective respondent was contacted personally and the information elicited from them. It allowed for elaborations to be made on some questions to facilitate easy responses. Probes were introduced where necessary during the interview to help ease their responses. Questionnaires were administered to property-owners Bawku Highways, to solicit their views on the factors identified from literature to have motivated the haphazard siting of buildings. The interviews were used to collect data from the planning authorities and the researcher had a discussion with the Assemblyman within Bawku Highways.

3.6 Methods of data analysis and reporting

The primary data were compiled, coded and entered into the computer for further processing and analysis. Quantitative techniques were used to report the findings from the questionnaire; thus, the Statistical Package for the Social Sciences (SPSS) was extensively used in this regard and Qualitative techniques were used to report the findings from the interviews and the discussion between the Assemblyman and the researcher. The output summaries were reformatted into tables, frequency, percentages; cross tabulation, charts and graphs were also used to communicate some of the findings.

4.0 RESULTS AND DISCUSSION

4.1 Introduction

This section examines the data obtained from the field in line with the objective of the paper. Results are presented and analyzed to show interrelationships of variables and to discover trends and patterns. A largely descriptive analysis is undertaken using tabulations with frequencies and simple percentages.

4.2 Demographic Attributes

The study explored the demographic characteristics of the respondents to enable further investigation as to the relationship between the demographic features and haphazard property development in the study area. Property development is undertaken by people of different age brackets, gender and education status. The analysis of the demographic characteristics indicated that out of a total of 120 respondents, 20 (16.67%) were females and 100 (83.33%) were males **(Table 1).** The higher (83.33%) male against the 16.67% female representation in the sample size was to reflect the voice of males telling their own stories as to how as to their motivation for the haphazard property development. This was very appropriate for the study because the Bawku Municipality is practicing the patrilineal system and consequently nearly all economic activities are dominated by men. Housing needs in the community is customarily the responsibility of the men.

The study also comprised 24(20.0%) of the respondents between the ages of 20 and 30 years, 54 (45.0%) of the respondents between the ages of 31 and 45 years. The remaining 42 (35.0%) of the respondents between the ages of 46 and above years **(Table 2).** This is a true reflection of the economic dynamics of the Municipality because people within the ages of 20 and 30 years are within the active school going ages and consequently do not have enough funds to put up buildings. If it does happen, it is an extreme case. And besides, the community is a low income earning class and dominated by agriculture as the main occupation. This reflects in the weak financial strength of most the inhabitants at their early years. In additions, it has become a normative compliance that people only start to put houses if they have plans of separating themselves from the extended family or getting married and these happen at the later stage of their lives and consequently reflects for the little representation in the sample size.

The analysis of the educational status of the respondents reveals that majority of the respondents had at least basic education 80(66.67%) and 40(33.33%) had no formal education at all (**Table 3**). This certainly prompted for the need to investigate the relationship between the education status and the rest of the variables under investigation.

Table 1: Gender of the respondents

Gender	**No. of Respondents** Total=120	**% of the Respondents** Total=100%
Male	100	83.33
Female	20	16.67

Source: Researcher's Field Survey, 2016.

Table 2: Ages of the respondents

Age Bracket (years)	**No. of Respondents** **Total=120**	**% of the Respondents** **Total=100%**
20-30	24	20.00
31-45	54	45.00
>45	42	35.00

Source: Researcher's Field Survey, 2016.

Table 3: Educational status of the Respondent

Educational Attainment	**No. of Respondents** **Total=120**	**% of the Respondents** **Total=100%**
No formal education	40	33.33
Basic education	25	20.83
Secondary education	35	29.17
Tertiary and Degree education	20	16.67

Source: Researcher's Field Survey, 2016

4.3 Awareness of the need for Planning Permission for property development

The haphazard property development in the Bawku Municipality was investigated by unfolding the awareness level of the inhabitants of the need to acquire a planning permit, development permit and an environmental permit for property development. The respondents were asked to indicate whether or not they are aware of the need to obtain planning documentation before initiating property development. Of the 120 respondents, 30 (25%) of the respondents indicated that they were aware of the need and the remaining 90 (75%) of the respondents **(Table 4)** indicated they were not aware of the need for planning documentation before property development. This certainly reflects general low level of education in the municipality or perhaps the failure of the planning authorities to create the awareness in that regard.

Of the (30) 25% of the respondents who indicated they were aware of the planning permissions, they were further asked as to indicate which of the planning permissions they were aware of. The analysis revealed that 10 out of the 30 respondents were aware of only planning permit, 15 out of the 30 indicated they were aware of both planning and development permit and 5 of the 30 indicated they were aware of planning, development and environmental permit **(Table 5).** This is certainly not a healthy indication as only fewer people will incorporate environmental and biodiversity conservation in their property development. The planning authorities and law enforcement agencies are to be blamed for this level of planning documentation awareness.

Table 4: Level of planning documentation awareness for property development

Are you aware of any planning permission for property development?	**No. of Respondents Total=120**	**% of the Respondents Total=100%**
Yes	30	25.00
No	90	70.00

Source: Researcher's Field Survey, 2016.

Table 5: Awareness level based on Category of planning permission

Type of Planning Permit	No. of Respondents Total=30	% of the Respondents Total=100%
Planning Permit Only	10	33.33
Planning and Development Permit only	15	50.00
Planning, Development and Environmental Permit	5	16.67

Source: Researcher's Field Survey, 2016.

The study further then investigated the motivations for the haphazard property development in the Bawku Municipality based on the Respondents' perspective and stakeholder's perspective as can be seen below.

4.4 Motivations for Haphazard Property Development *(property owners' Perspective)*

The nucleus of the paper was to make a paradigm shift from conceptual framework into empirical studies to unfold the actual motivations for the haphazard property development in the Bawku Municipality. From the perspective of the respondents a number of practical motivations were identified **(Table 6).** The indications from the property owners were given much attention because they are the major players in the haphazard property development.

Table 6: Motivations from the property owners' perspective

Motivations	No. of Respondents
Ignorance on planning and building regulations	**90 (75.00%)**
Nature of the Property	**80 (66.67%)**
High cost of obtaining land document and building permits	**30 (25.00%)**
Unrealistic zoning	**18 (15.00%)**
Non punitive sanctions against offenders	**40 (33.33%)**
Bureaucracy	**15 (12.50%)**
High cost of land	**15 (12.50%)**
Location of land	**10 (8.33%)**
corruption	**68 (56.67%)**

Source: Researcher's Field Survey, 2016

The implications or indication from the table above are that, the most endemic motivation was unawareness of planning and building regulations 90(75.00%). The respondents stated that they did not obtain building permits before constructing their properties because they were not aware of the obligatory planning need to obtain a building permit before siting their houses. This is very consistent with the number of respondents who were unaware of the need for planning permits to be obtained before initiating development of properties. An old man who was a respondent expressed this customarily as:

> *"the land is my own and do I have to ask someone for permission before I start building.....my grandfather gave me this later and he is no more and therefore I am the outright owner of the land....i can use it for anything without restrictions.......(A., a property owner)"*

Nature of the property 80(66.67%) was one of greatest motivation for the refusal to acquire planning permissions and consequently poor zoning of such property by the owner. Until recently, the most dominant type of residential buildings were basically wattle and daub ('Atakpaame' in the Ghanaian parlance) carefully build of laterite and roofed with thatch. Even now, one can identify a number of viable lands dominated by round houses built of laterite. These property owners found it too much of a stress acquiring planning permit for such properties. A vegetable retail shop owner expressed this clearly when she said

> *"my shop is not permanent and it is even too small for me to acquire planning permission before setting it up........besides, the permit's fee is far more than the total cost and revenue to be obtained from my shop for two years.....(L.; a shop owner)"*

Another serious motivation that was identified by some of the property owners was the corruption of the planning officials 68 (56.67%). They indicated that if you submit your building particulars for building permits, the planning authorities demanded other fees or otherwise, the request was delayed. This obviously reflects the occurrence of poorly sited buildings even though some of the home owners 30(25.00%) are aware of planning permits.

Some of the property owners indicated that no authority administers any form of punishment to those who violates the planning requirement and hence identified no punitive sanctions against offenders 40 (33.33%) as their motivation. One of the property owners expressed this in the following words:

> *"my neighbours started buildings very close to the road and I thought any authority will come to stop him but he completed the buildings and is living inside the house with his family.....i reported him to the assemblyman who also reported him to the planning authorities but they have not said anything....(M.; a property owner)"*

It was also identified that high cost of obtaining land document and buildings permits 30(25.00%) was a motivation for the unauthorized siting of properties in the Bawku municipality. They indicated that they would rather poorly site their buildings than to pay high fees for buildings permit in the Municipality. That was a great sign of ignorance because these people are not aware that the effects of their haphazard development will visit them in several directions, ways and magnitude.

Other motivations for the unauthorized siting of buildings in the Bawku Municipality from the property owners' perspective included unrealistic zoning 18 (15.00%), bureaucracy 15 (12.50%), location of land 15 (12.50%) and high cost of land 10 (8.33%).

4.5 Motivations for Haphazard Property Development *(Stakeholder's Perspective)*

The study diversified the investigation to also include the views of stakeholders regarding the motivations for the unauthorized siting of properties in the Bawku Municipality. The stakeholders in this regard included the planning authorities the Bawku Municipality and the assemblyman of the Bawku Highways. The following motivations were identified and ranked as follows: **(Table 7).**

Table 7: Motivations from the stakeholders' perspective

Motivation	**Rank**
Indiscipline of the property owners	**1st**
Population growth and unregulated urbanization	**2nd**
Logistic and capacity gap	**3rd**
Lack of political will to enforce building regulations	**4th**
High enforcement cost	**5th**
Lack of public-private partnership in controlling unauthorized buildings	**6th**

Source: Researcher's Field Survey, 2016.

From the table above, a group discussion and interview with planning authorities of the Bawku Municipality and the assemblyman of the Bawku Highways indicated a number of motivations for the unauthorized siting of properties in the Municipality. The stakeholders identified and ranked the indiscipline of the properties owners as the 1st and most practical cause. This presupposes that, the planning authorities believe the property owners are aware of the need of the building permits and yet will not for any reason acquire them. This contradicts with the findings that most of the property owners 90 (75.00%) were not aware of the obligatory need for planning in property development. The stakeholders however identified and ranked lack of public-private partnership in controlling unauthorized buildings 6th as a motivation. By this, they believe the stakeholders are not helping them in enforcing the planning and building regulations.

The other motivations were identified and ranked from 2nd to 5th as follows: population growth and unregulated urbanization, logistic and capacity gap, lack of political will to enforce building regulations and high enforcement cost.

4.6 Effects of Haphazard Property Development

The haphazard or unauthorized property development has a number of effects on the communities concerned.

In the first place, the unauthorized siting of the properties presents a poor planning outlook of the Municipality. One cannot describe a particular area as a residential neigbourhood or a commercial neigbourhood. It is very common to locate highly incompatible land uses situated in a particular neigbourhood. The properties are developed everywhere and anywhere. This gives a very discomforting outlook to the Municipality.

Another effect of the haphazard property development is threat to future infrastructural development. Public water mains and utilities do not find suitable paths for installation and the result has been that most of the buildings do not receive the public services. Nearly all lands earmarked for future development are been used for less useful land uses. It is very common to identify houses in buffer zones for road expansions. Lands that have been allocated for schools, markets, open spaces, forest reserve, among others have been colonized and used for compound houses.

A very serious threat to the health of the inhabitants is also a consequence of the unauthorized property development in the Municipality. One will notice people putting houses in water courses, area liable to flood, marshy areas and poor sanitation areas. In cases of natural disasters, the casualties would undesirable. In times of heavy downpours, people houses are sometimes submerged in the water and eventually health related problems are always on the increase during the rainy season.

Another effect has been that such development often threatens the natural environment. Because environmental permits are not often obtained before initiating such property development, the property owners do not interweave the buildings with the natural environment and the result has been loss of biodiversity and rare plant species.

Finally, waste of land is one of the effects of unauthorized siting of properties. Because such developments do not conform to zoning regulations, buildings are often placed in a manner that adjacent lands becomes waste and cannot be used for economic development.

4.7 Indication for the Planning Authorities and Stakeholders

The findings of the study regarding the motivations and effects of haphazard property development dictate that planning authorities increase their attention to the regulation of property development. Even though the Municipality has been bedeviled by chieftaincy disputes for five decades (1959-2009), peace has returned to the municipality and hence the enforcement of these regulations is highly possible and feasible. The planning authorities need to increase their human resources base and awareness campaigns. These should be sustained. The stakeholders of the municipality also have a role to play in this regard.

5.0 Conclusion and Recommendations

It is very safe to conclude that both the stakeholders and the property owners are the actual causes of the unauthorized property development in the Municipality and accordingly, the study posits the following recommendations as practical ways correcting the state of the art or minimizing the spread of such developments and the negative impact as well.

The study recommends that the planning authorities and stakeholders should improvise and sustain mechanisms for creating awareness and educating the inhabitants on the constitutional obligatory need to obtain planning permissions before commencing property development. They should be educated on the importance of respecting and obeying zoning regulations and conserving the natural environment.

The study also recommends that after, the creation of the awareness, people found guilty should be given punitive sanctions including demolition of such development at the earliest stage of development, conviction or conviction with fines to discourage people from such practices. In this regard justice must not be tempered with mercy to protect the municipality from urban sprawl.

It is also recommended that, the process of obtaining the building or planning permits should be made less expensive and bureaucratic should be practically eliminated to galvanize people to patronize the process.

REFERENCES

- Acquah-Harrison, R. (2004). *Housing and Urban Development in Ghana: With Special Reference to Low-Income Housing*. United Nations Human Settlement Programme, Nairobi.
- Adams, A. (2012). *Building on Waterways Assumes Alarming Proportions in Sekondi-Takoradi*. The Chronicle, 31 May 2012.
- Adams, I. (2014). *The Development Dilemma: Give a Man a Fish or Teach Him to Fish?* The Guardian, 16 January 2014. http://www.theguardian.com/global-development-professionals-network/2014/jan/16/agbogbloshie-ghana-global-com
- Adjei Mensah, C. (2010). *Causes and Consequences of Informal Settlement Planning in Ghana: A Case Study of Aboabo, a Suburb of Kumasi Metropolis. M. Phil. Thesis*, University of Cape Coast, Cape Coast.
- Aklorbortu, M.D. (2013). *Rush for Land Compromises Flight Safety in Takoradi*. Daily Graphic. http://graphic.com.gh/archive/General-News/rush-for-land-compromises-flight-safety-in takoradi.html
- Ali, M.H. and Sulaiman, M.S. (2006). *The Causes and Consequences of The Informal Settlements in Zanzibar. Proceedings of the XXIII FIG Congress,* Munich, 8-13 October 2006, 1/17-15/17.
- Badu, E., Owusu-Manu, D., Edwards, J.D., Adesi, M. and Lichtenstein, S. (2013). *Rural Infrastructure Development in the Volta Region of Ghana: Barriers and Interventions*. Journal of Financial Management of Property and Construction, 18, 142-159. http://dx.doi.org/10.1108/JFMPC-11-2012-0040
- Cheema, G. (1993). *The Challenge of Urban Management: Some Issues. In: Cheema, G. and Ward, S., Eds., Urban Management Policies and Innovations in Developing Countries*. Praeger Westport, London, 17.
- Creswell, J. (2007). *Qualitative Inquiry and Research Design: Choosing among Five Approaches*. Sage, Newbury.
- Freiku, S.R. (2003). *Kumasi Developments Getting Worse, Chaotic*. Ghanaian Chronicle, 31 January 2003. http://www.modernghana.com/news/30669/1/kumasis-development-getting-worse-chaotic.html
- Ioannidis, C., Psaltis, C. and Potsiou, C. (2007). *Towards a Strategy for Suburban Informal Building Control through Automatic Change Detection. Arbitrary and National Economy, the Need of the Real Estate Market for Modern Cadastre and Spatial Planning, TEE, ATM, WPLA, FIG: 28-31 March, 2007*, Cape Sounion.
- Kings-Amadi, I. (2004). *Appraisal of Land-Use Plan Implementation in River State: A Case Study of Port Harcourt Metropolis. Unpublished Master's Thesis*, Rivers State University of Science and Technology, Institute of Geosciences and Space Technology, Port Harcourt.
- Kumar, P. (2012). *Unauthorized Constructions and Earthquake Disaster Impact Mitigation. ISET Golden Jubilee Symposium, Roorkee, 20-21 October 2012, Paper No. G003:1-7.*
- Magigi, W. and Majani, B.B.K. (2006). *Housing Themselves in Informal Settlement: A Challenge to Community Growth Processes, Land Vulnerability and Poverty Reduction in Tanzania. Proceedings of the 5th FIG Regional Conference, Accra, 8-11 March 2006, 1/24-24/24.*
- Melesse, M. (2006). *City Expansion, Squatter Settlements and Policy Implications in Addis Ababa: The Case of Kolfe Keranio Sub-City, Addis Ababa*. Department of Geography and Environmental Studies, Addis Ababa University, Addis Ababa.
- Owusu-Mensah, K. (2003). *Community Participation in Town and Country Planning: A Delayed but Welcoming. GNA. http://www.modernghana.com/news/112686/1/community-participation-in-%20townplanning-a-delayed.html*

- Sietchiping, R. (2000). *Understanding Informal Urban Patterns by Integrating Geographic Information System (GIS) and Cellular Automata (CA) Modelling.* Department of Geography and Environmental Studies, Melbourne.
- Smith, M.P. (1980). *The City and Social Theory.* Blackwell, Oxford.
- Somiah, M.K., Osei-Poku, G. and Aidoo, I. (2015). *Relative Importance Analysis of Factors Influencing Unauthorized Siting of Residential Buildings in the Sekondi-Takoradi Metropolis of Ghana. Journal of Building Construction and Planning Research, 3, 117-126.* http://dx.doi.org/10.4236/jbcpr.2015.33012 *and Spatial Planning, TEE, ATM, WPLA, FIG: 28-31 March, 2007*, Cape Sounion.
- United Nations (2007). *Discussion Paper on Challenges and Integrated Policy Responses for Informal Settlements.* Economic and Social Council, Geneva.
- Yin, R.K. (2003*). Case Study Research: Design and Methods. 3rd Edition*, Sage, Thousand Oak.